Climate Change Implications for Irrigated Agriculture in the Lower Indus Basin. Wheat Production in Pakistan Provinces Sindh and Punjab

Merle Becker

Bibliographic information published by the German National Library:

The German National Library lists this publication in the National Bibliography; detailed bibliographic data are available on the Internet at http://dnb.dnb.de.

ISBN: 9783346756510
This book is also available as an ebook.

© GRIN Publishing GmbH
Nymphenburger Straße 86
80636 München

Print and binding: Books on Demand GmbH, Norderstedt, Germany
Printed on acid-free paper from responsible sources.

The present work has been carefully prepared. Nevertheless, authors and publishers do not incur liability for the correctness of information, notes, links and advice as well as any printing errors.

GRIN web shop: https://www.grin.com/document/1291517

Universität Hamburg

Institut für Geographie

Merle Becker

Report

Climate Change Implications for Irrigated Agriculture in the Lower Indus Basin: Wheat Production in Pakistan Provinces Sindh and Punjab

Wintersemester 2021/22

Course: 63-111 Climate Change Impacts on Water Resources of High Asia

Content

1 Introduction

Temperature increase, precipitation decrease or changing intensity, heat waves and droughts – climate change is predicted to affect the world's agriculture adversely. In line with other regions, South Asia will show agricultural productivity decrease, according to IPCC (cf. ABID 2016, p. 20). The Lower Indus Basin as part of the Indus Basin in Pakistan, however, is a special region since it has the largest irrigation system in the world (cf. ASGHAR et al. 2019, p. 137). This means that 80% of all arable land is irrigated and irrigation produces 90% of all food in this area. Therefore, temperature and precipitation are not the only physical-meteorological factors for agricultural production affected by climate change, but one must also consider the Indus River runoff since this is the main source for irrigation (cf. QURESHI 2011, pp. 252 f.). The Indus River is mainly dependent on snow and glacier melt in the Upper Indus Basin, they contribute about 80% to the total runoff (cf. ARCHER & FOWLER 2004, p. 47), so future runoff might be affected due to temperature and precipitation changes, which lead to accelerated glacier melt. Since Pakistan shows a major population growth, future water demand will be higher (cf. QURESHI 2011, p. 252), and important issues will be food security and water management.

Wheat is the most important food crop in Pakistan with a share of 60% to 80% of the total cropped area for food crop production (cf. AHMED & SCHMITZ 2011, p. 3), which is reflected in the population's food consumption (cf. IMRAN & NOUREEN 2021, p. 8). These observations arise the question of how wheat production will be affected by climate change in the future because it seems to be of high importance to Pakistan.

In the past years, this issue was investigated in several studies. IPCC points out that "positive and negative changes are mixed in simulated wheat and maize yields" (IPCC 2019, p. 454) on a global scale, though studies in the Lower Indus Basin (LIB) show contradictory results on a regional scale as well. It is noticeable that most studies concentrate on temperature and precipitation changes (e.g., AHMAD et al. 2014; LOHANO & MARI 2020a; MAHMOOD et al. 2021) as well as extreme weather events (e.g., ABBAS et al. 2018; SAEED et al. 2017) when making future projections for wheat production. These studies do not include the factors irrigation and river runoff accordingly. BIEMANS et al. (2019) investigate the timely water demand and availability from Indus runoff, but do not consider temperature and precipitation changes.

Therefore, this study aims to summarize literature findings on climate change in the LIB and its impacts on wheat production for the three factors (1) temperature and precipitation, (2) extreme weather, and (3) Indus River runoff, and to derive a general statement on these impacts for the future. The research question to be answered is the following: *Which implications for wheat production in the Lower Indus Basin can be derived from climate change and its impacts on Indus River runoff?* The research is restricted to the Pakistan provinces Sindh and Punjab, which are the most important regions for wheat production (cf. AHMED & SCHMITZ 2011, p. 3).

To answer the research question, the study area will be introduced firstly. Secondly, climate change induced developments of the three factors and their impacts on wheat production will be presented. The findings will be summarized to a general statement and future implications for adaptation strategies will be presented.

2 Study Area

Physiographic character

The provinces Sindh and Punjab cover most of the LIB's area and lie in Pakistan, which borders to Himalaya in the north and Arabian Sea in the south (cf. HABIB 2004, p. 34). The Indus Basin is a transboundary region: the Upper Indus Basin (UIB) stretches from Afghanistan to India, which can be seen in fig. 1. The Indus River originates from several tributaries in the high elevated areas of the UIB in the Himalayan mountains and flows through the low elevated LIB into Arabian Sea (cf. POMEE & HERTIG 2021, p. 4). Being the main river in the basin, most of the other smaller rivers in the area flow into Indus River (cf. AHMAD et al. 2012, p. 5256).

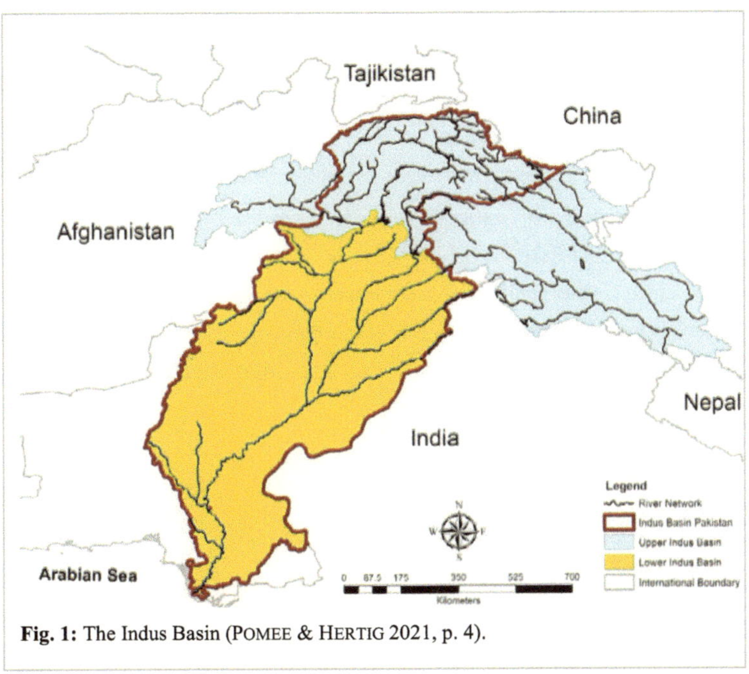

Fig. 1: The Indus Basin (POMEE & HERTIG 2021, p. 4).

From the 1950s, several water projects were implemented along the Indus River with the aim of flood control and water transfer to a canal network for irrigation. Before these measures, the river had meandered and flooded large areas seasonally, or canal avulsions had delivered water to lower areas in the plains. The construction of dams reduced flood peaks and made it possible to distribute water efficiently into the lower plains (cf. JORGENSEN et al. 1993, p. 294).

Regarding climate, there are big differences within the province Punjab because of the high latitudinal extension and the impact of the mountain ranges in the north. The climate in Sindh and southern Punjab is arid with hot summers and mild winters, whereas northern Punjab is semiarid. In Sindh, summer extends from March to November and most months' mean temperature is above 30°C. In winter, temperature ranges between 10°C and 21°C, and frost may occur on very cold days. The annual rainfall is below 250mm in most areas and concentrates on summer monsoon. Southern Punjab, in comparison, shows a little lower mean temperatures and a shorter summer as well as more precipitation of 250mm to 375mm (cf. KHAN 1991, pp. 43–45). Northern Punjab shows similar temperatures like the southern part, but the main difference lies in the precipitation, which ranges from 325mm to 755mm in the most northern parts (cf. ibid., p. 46 f.). The climatic differences indicate that conditions for wheat production differ within in the provinces.

Agriculture

Pakistan depends highly on the agricultural sector; it makes up 21% of the country's GDP. Together with agro-based products, agriculture accounts for 80% of the export earnings, and more than 43% of labor force is engaged in this sector. Crops are the most important agricultural branch, contributing 40% to Pakistan's agriculture (cf. IMRAN & NOUREEN 2021, p. 8). Being the most important food crop in Pakistan, wheat makes up 60% to 80% of the total cropped area, and Sindh and Punjab are especially suitable for wheat production (cf. AHMED & SCHMITZ 2011, p. 3). The cultivation period for wheat is November (sowing) until March / April (harvest), meaning that wheat is a winter crop (cf. MAHMOOD et al. 2021, p. 2926).

Fig. 2 shows the irrigated areas in both provinces. Since there are several major tributaries additionally to Indus River in Punjab, the irrigated area covers almost the whole province, whereas there is only the Indus River (and minor rivers) in Sindh, so irrigation is located around Indus River there. Therefore, Punjab is the most important region for crop cultivation. The total crop area is 16.7 million ha, which is more than five times the total crop area in Sindh (3.1 million ha). Punjab's agriculture contributes 80% to the total wheat output, Sindh 12% (cf. MAHMOOD 2009, p. 2924). However, wheat yield is higher in Sindh (3200kg/ha) than in Punjab (2800kg/ha) (cf. KHAN et al. 2021, p. 8000).

After introducing the study area, the next aim is to summarize climate change impacts that might affect wheat production in both provinces.

3 Climate Change in the Indus Basin

To assess climate change impacts on wheat production in the LIB, it is necessary to first determine which indicators affect agriculture and how they interrelate with climate quantities. Wheat crops have a certain temperature and water demand, and these two main factors will be investigated.

PORTER & GAWITH (1999) state that wheat growth shows its optimum between 17°C and 23°C and is stopped below 0°C and above 37°C (p. 25), however, IMRAN & NOUREEN (2021) refer to a temperature of 30°C to 32°C above which wheat growth stops, an optimum of 25°C and a minimum of 0 to 5°C (p. 15). Hence, temperature changes must be investigated with respect to optimum values as well as changing temperature extremes due to wheat growth stop.

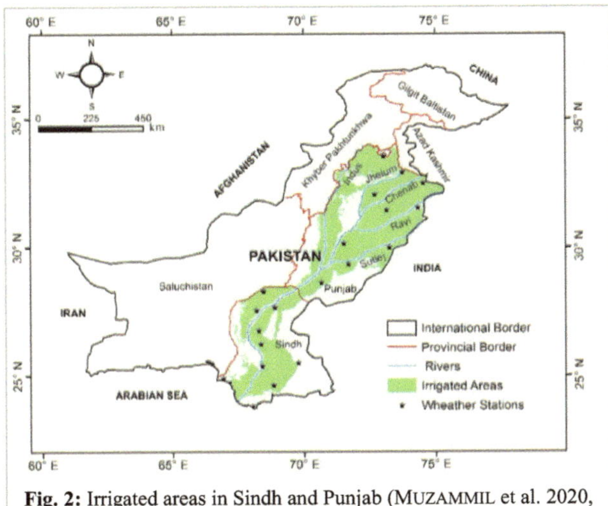

Fig. 2: Irrigated areas in Sindh and Punjab (MUZAMMIL et al. 2020, p. 3).

Water sources are precipitation and irrigation. Especially Sindh is highly dependent on irrigation from the Indus River since precipitation in winter is below 5mm in all months (cf. World Bank Group 2021). However, water availability from Indus River is limited in winter as well and wheat is mainly irrigated with groundwater (cf. BIEMANS et al. 2019, p. 600) because the runoff is mostly fed by snow and glacier melt in the UIB, which show their peak in spring and summer (cf. ibid., p. 596). If the winter runoff increased in the future, more water would be available to irrigate a larger region and thus wheat production would increase. On the other hand, if the runoff decreased, crop area would decrease as well (cf. AHMAD et al. 2020, p. 32).

Since Indus River runoff depends on hydrological processes in the UIB, it is necessary to investigate climate change impacts on them and the indications for the future runoff. Precipitation changes must be considered as well since a major increase in winter would have positive impacts on wheat production due to more water available. However, at least two studies conclude that precipitation changes do not have a significant impact on wheat productivity because firstly, precipitation changes will be non-significant themselves, and secondly, other water sources are used for irrigation (MAHMOOD et al. 2021; AHMAD et al. 2014). These observations lead to three main factors that this study aims to investigate:

(1) Mean temperature

(2) Extreme temperature patterns

(3) Indus River runoff

Other factors like, e.g., floods, pollution, or soil salinity are not considered in this study.

Next, different literature findings on these factors are summarized. Since past changes often correspond to future changes, both will be analyzed. Literature findings differ from each other for the following reasons: different time periods or reference periods used, different study areas (which differ in their climate due to latitude or altitude) as well as different methods (for calculation models, e.g.) or different climate scenarios and simulations. Most studies refer to the IPCC's RCP scenarios, mainly RCP 4.5 (intermediate emissions) and RCP 8.5 (high emissions), though some do not. These scenarios determine certain scenarios of greenhouse gas concentrations including socio-economic aspects (cf. KASANG n. d.). The aim now is to derive trend ranges for different future scenarios from selected literature.

3.1 Mean Temperature

Past changes

For the past centuries, all studies find an increasing trend of temperature for the LIB. Tab. 1 gives an overview on the findings. The values differ significantly within the region, which is shown especially by SAJJAD & GHAFFAR (2019) and SHAFQAT et al. (2016), who use meteorological data from certain places in the provinces. Also, temperature increase has been significantly higher in Sindh than in Punjab, which is shown by World Bank Group (2021) and SAJJAD & GHAFFAR (2019), with a major increase in winter (SHAFQAT et al. 2016) or spring (S. KHAN et al. 2019), respectively. For the period 1960 to 2021, a mean temperature increase of +1.4°C (Sindh) and +0.9°C (Punjab) can be calculated from the values from S. KHAN et al. (2019), SAJJAD & GHAFFAR (2019), World Bank Group (2021), and SHAFQAT et al. (2016).

Future trends

In the future, past trends are most likely to continue, e.g., temperature increase in winter will be higher than in summer like in the past. However, the trend for Sindh and Punjab will reverse, meaning that temperature increase will be higher in Punjab in the future. The difference will be around 0.1°C to 0.2°C for mid-century (cf. SAJJAD & GHAFFAR 2019; World Bank Group 2021) and 0.3°C to 0.5°C for late-century (cf. World Bank Group 2021). This is an important finding

Study	Time period	Study area	Findings
Lohano & Mari (2020)	1960 - 2019	Sindh	Most meteorological stations show an upward trend (only qualitative)
Lutz & Immerzeel (2016)	1981 - 2010	LIB	+ 0.9°C
S. Khan et al. (2019)	1960 - 2013	Sindh & Punjab	Increase of mean maximum and minimum temperature: Sindh: + 0.22°C/decade (max), + 0.25°C/decade (min) Punjab: + 0.23°/decade (min and max) Highest increase during spring time: + 0.34°C/decade (max, Sindh) and +0.4°C/decade (max, Punjab)
Sajjad & Ghaffar (2019)	1960 - 2013	Sindh & Punjab	Increase of mean maximum and minimum temperature: Sindh: + 1.6°C (max), +1.6 to 2.2°C (min) Punjab: + 1°C (max), + 1.1 to 2.6°C (min)
Shafqat et al. (2016)	1953 - 2006	Punjab	Difference of mean temperature in 1953-1979 and 1980-2006: Lahore (Punjab): + 0.9°C ; Multan (Punjab): + 0.23°C Highest increase in winter, little decrease in summer
World Bank (2021)	1960 - 2020	Sindh & Punjab	Sindh: + 1.3°C Punjab: + 0.44°C

Tab. 1: Summary of literature findings on past temperature trends in the LIB (own depiction).

since Punjab is the most important region for wheat production. The greater impact from temperature increase is therefore of high importance for wheat production in whole Pakistan. However, one must keep in mind that Punjab's climate is cooler than Sindh's. The annual cycle of temperature will not be affected (cf. RAJBHANDARI et al. 2015, p. 349) apart from higher temperature variations due to the seasonal difference in temperature increase (cf. LOHANO & MARI 2020a, p. 670).

To compare the findings for future changes (see Tab. 2), it is important to have as many equal parameters as possible. Thus, only values for mid- and late-century are listed to provide the same time period. YU et al. (2013) do not use RCP scenarios, but the A1B and A2 scenario. As the A2 scenario corresponds to RCP 8.5, their findings are shown here. RAJBHANDARI et al. (2015) use PRECIS simulations, which stand for different scenarios as well. The range of temperature rise for three of these scenarios is given in Tab. 2. One barrier to the comparison of results is that different reference periods were used. As shown before, within these periods there has been a temperature increase of about 1°C, whereas the major increase took place between 1990 and 2010 (cf. World Bank Group 2021), so more recent reference periods should lead to lower values for future temperature increase. In fact, both RAJBHANDARI et al. (2015) and YU et al. (2013) find higher values for mid-century using the reference period 1961 – 1990 than LOHANO & MARI (2020a), SAJJAD & GHAFFAR (2019) and World Bank Group (2021), whose reference periods reach into 21st century. For this reason, mean values are calculated in Tab.3 from the studies with recent reference periods only. The values are comparable even though the periods differ because they all include the time period with the major temperature increase in the past. Since not all three studies present values for all scenarios shown here, some mean values depict only two studies. World Bank Group (2021) is the only source that gives values for RCP 4.5 scenario and late-century.

Tab. 3 shows the calculated mean values for a reference period between 1960 and 2014. It

Study	Time period	Study area	Findings
Lohano & Mari (2020)	2016 - 2035 & end of 21st century	Sindh	Reference period: 1960 - 2007 **RCP 4.5:** Dec-Feb: + 0.75°C to + 1.5°C Jun-Aug: + 0.5°C to 1°C **RCP 8.5:** + 4°C to 5°C (end of 21st century)
Rajbhandari et al. (2015)	2020 - 2070	LIB	PRECIS simulations, reference period: 1961-1990 **2041-2070:** + 2.5°C to 3.5°C; **2071-2098:** + 3.5°C to 6°C Higher increases in winter than in summer
Sajjad & Ghaffar (2019)	2045	Sindh & Punjab	Reference period: 1975 - 2013, mean maximum and minimum temperature **RCP 4.5:** Sindh: + 0.8°C (max), + 1°C (min) Punjab: + 1°C (max and min) **RCP 8.5:** Punjab: + 1.1°C (max), + 1.4°C (min) Sindh: + 1°C (max), + 1.2°C (min)
Yu et al. (2013)	2050 & 2080	Southern Pakistan	Reference period: 1961-1990 GCISC model, **A2** scenario (≈ RCP 8.5) **2050:** + 2.4°C ; **2080:** + 4.2°C Winter increase higher than in summer
World Bank Group (2021)	2040 - 2059 & 2080 - 2099	Sindh & Punjab	Reference period: 1995 - 2014 **2040 - 2059:** Sindh: + 1.36°C (RCP 4.5) ; + 1.8°C (RCP 8.5) Punjab: + 1.42°C (RCP 4.5) ; + 2°C (RCP 8.5) **2080 - 2099:** Sindh: + 2.3°C (RCP 4.5) ; + 4.4°C (RCP 8.5) Punjab: + 2.6°C (RCP 4.5) ; + 4.9°C (RCP 8.5) Winter increase higher than in summer

Tab. 2: Summary of literature findings on future temperature trends in the LIB (own depiction).

becomes clear that the found differences between Sindh and Punjab increase with time and that there is a great difference between the two scenarios especially in late-century. Seasonal data was not considered here, so these mean values refer to an annual scale. Since winter temperature increase is projected to be higher than in summer, it can be expected that winter temperature increase values lie above the calculated values in Tab. 3.

		RCP 4.5	RCP 8.5
Mid-century	Sindh	+ 1.1°C	+ 1.5°C
	Punjab	+ 1.14°C	+ 1.6°C
Late-century	Sindh	+ 2.3°C	+ 4.5°C
	Punjab	+ 2.6°C	+ 4.8°C

Tab. 3: Calculated mean values (own depiction).

3.2 Extreme Temperature Patterns

As outlined at the beginning of chapter 3, wheat is sensitive to extreme temperatures at both ends of the temperature scale. Therefore, main climate factors are the number of heat waves and hot days as well as the number of cold or frost days. Other extremes like consecutive dry or wet days etc. are not included in the analyses of climate change impacts on wheat growing and will therefore not be part of this study either.

N. KHAN et al. (2019) define heat wave as the "[n]umber of consecutive days with temperature > 95th percentile of the daily maximum temperature" (p. 904) and hot days as the "[n]umber of days having temperature > 95th percentile of the daily maximum temperature" (ibid.). Other authors use different definitions, e.g., they use the 90th percentile (e.g., ABBAS et al. 2018), which leads to differing results. Five studies are presented here. A quantitative analysis and comparison of literature findings is beyond the scope of this paper, so the qualitative trends are summarized in Tab. 4.

For the past, ABBAS et al. (2018), N. KHAN et al. (2019) and SAEED et al. (2017) find an increasing trend for both hot days and extreme cold days as well as heat waves, whereas the increase is found to be higher either in Punjab (cf. SAEED et al. 2017, p. 1641) or in Sindh (cf. N. KHAN et al. 2019, p. 910). For the future, the trend in increasing heat waves and hot days will continue, also in winter (cf. SAEED et al. 2017, p. 1644), which is relevant for wheat production. On the other hand, the number of cold days and frost days will show a reversing, thus decreasing trend, as the findings of SAJJAD & GHAFFAR (2019) and World Bank Group (2021) show.

Study	Time period	Study area	Findings
Abbas et al. (2018)	1980 - 2015	Sindh (Indus Delta)	Number of hot days increased Number of coldest days increased
N. Khan et al. (2019)	1960 - 2013	Sindh & Punjab	Number of heat waves and hot days increased, especially in Sindh Number of cold waves decreased, number of extreme cold days increased
Saeed et al. (2017)	1979 - 2000 & 2026 - 2100	Sindh & Punjab	**Past:** Number of heat waves increased, especially in upper Sindh and Punjab **Future:** Number of winter heat waves will increase throughout the whole century, perhaps more in Punjab
Sajjad & Ghaffar (2019)	2016 - 2045	Sindh & Punjab	Number of cold days will decrease for both provinces
World Bank Group (2021)	2040 - 2099	Sindh & Punjab	Number of very hot days (T>35°C) will increase in autumn and spring; especially in Sindh Number of frost days will decrease in winter

Tab. 4: Summary of literature findings on past and future trends in extreme temperature (own depiction).

Not all studies present seasonal data, though it is important for wheat production whether the trends in extreme temperature patterns will occur in the cultivation period. SAEED et al. (2017) hint that heat waves will increase in winter as well as in summer, and World Bank Group (2021) shows that the number of very hot days will show its major increase in autumn and spring, when wheat is sown or harvested. Fig. 3 shows the total number of very hot days (T>35°C) for mid-century and RCP 4.5 scenario in Sindh and Punjab. Sindh will show a high number of very hot days (7 to 25) in November and March / April, the edges of the cultivation period. The values for Punjab are a little lower. With respect to changes, major increase will be mainly in October / November and March / April by about 2 to 5 days for both provinces (cf. ibid.). From December to January, there is no or a non-significant increase in very hot days, though, it can be concluded that hot days with temperature above 30°C might as well increase in all months. For late-century, the increase of very hot days in Sindh will be around 7 to 8 days in November and March, and also February will show a significant increase. For RCP 8.5 scenario, the increase will be higher than for RCP 4.5 scenario (cf. ibid.).

The number of frost days is already very low in both provinces. For Punjab, the number of frost days will decrease from an annual value of 3.2 days to 0.9 days for mid-century and RCP 4.5, for Sindh from 0.3 days to 0.1 days (reference period: 1995 – 2014) (cf. World Bank Group 2021). This corresponds to the finding of a decreasing trend in cold days from SAJJAD & GHAFFAR (2019).

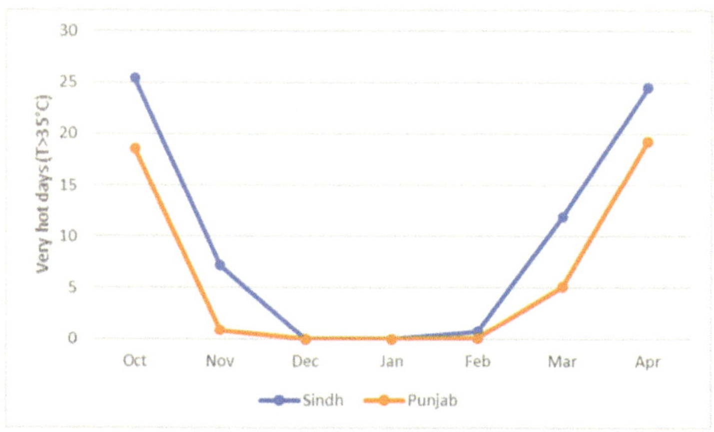

Fig. 3: Number of very hot days (T>35°C) for Sindh and Punjab, 2040-2059, RCP 4.5. Data from World Bank Group (2021) (own depiction).

3.3 Indus River Runoff

The Indus River is fed by seasonal and permanent snowfields and glaciers by more than 80%, monsoon rainfalls contribute only a small share. This means that summer runoff is highly correlated with preceding precipitation and temperature in the UIB where the Indus River originates (cf. ARCHER & FOWLER 2004, p. 47), and runoff shows a peak in summer between June and September like the tributaries from UIB do as well (cf. AHMAD et al. 2012, p. 5265). Therefore, temperature and precipitation changes in the UIB will determine future runoff changes. For the past, ALI et al. (2015) find a strong increase in winter temperature in the UIB (p. 1) and AHMAD et al. (2012) hint that winter precipitation shows a significantly increasing trend as well (p. 5259). Additionally, they investigate runoff changes of eight tributaries in the UIB and find that six of them have shown an increasing trend in the past centuries (cf. ibid., p. 5262). Thus, an increasing trend in Indus River runoff in winter should be found.

Past changes

When investigating Indus River runoff, most studies analyze data from tributaries in the UIB, which show different trends depending on their altitude. ARFAN et al. (2019) find increasing trends in Indus River runoff in winter and spring between 1961 and 2010 (pp. 9 – 11). KHATTAK et al. (2011) investigate runoff changes in the UIB on a spatial and seasonal scale. For upper regions, they find an annual increasing trend with major increases in autumn, winter, and spring. For middle regions, the annual trend is slightly decreasing, though showing a strong increase in winter and spring. Lower regions, in contrast, show a decreasing trend in all seasons, but mainly in summer (cf. KHATTAK et al. 2011, p. 114). These runoff changes are attributed to temperature and precipitation increase especially for upper areas (cf. ibid., p. 116). For lower areas, there is a decrease in snowfall due to higher temperature, thus snowmelt runoff decreases (cf. NEPAL & SHRESTHA 2015, p. 213), and earlier snowmelt results in a runoff shift from spring and summer to winter, which leads to a winter runoff increase (cf. LUTZ et al. 2016).

Future trends

Findings on future runoff trends vary strongly because there are high uncertainties due to the interplay of temperature and precipitation, which themselves are uncertain (cf. ibid.). Again, different studies use different methods, study areas and time periods, leading to differing trends and values, but here the selected three studies use similar reference periods. GEBRE & LUDWIG (2014) investigate runoff changes in the Indus River outlet in Sindh, whereas LUTZ et al. (2016) gather data for several tributaries in the UIB and ALI et al. (2015) use data from one gauging station at Indus River in the UIB. Close to the outlet, snow and glacier melt contribute 60% to 70% to the Indus River runoff, in the upper areas 80% to 90%. Several main tributaries flow into Indus River in southern Punjab (e.g., Chenab, and Jhelum), which originate in eastern parts of the UIB (cf. BIEMANS et al. 2019, p. 597). These lower-latitude rivers are mainly fed by rainfall during monsoon season (cf. LUTZ et al. 2016), so this might be a reason for GEBRE & LUDWIG (2014) finding the strongest runoff increase in summer for the river outlet, whilst the other two find decreasing trends in summer for upper stations.

Tab. 5 shows the summary of findings for these three studies. The range of runoff changes were read from diagrams or tables for the cultivation period (November to March) and both RCP scenarios. Again, both scenarios show different results throughout 21st century. For lower Indus River, RCP 4.5 scenario shows similar increase for 2030 and 2070 of 20% to 80% with a little higher increase in 2070. RCP 8.5 shows a runoff increase that is significantly higher from 70% to 150%, again with little higher increase for 2070. For upper Indus River, the two studies' results show significant differences. LUTZ et al. (2016) find a weak increase up to 25% until mid-century and RCP 4.5, and for late-century a weak increase in some months, for others light decrease. ALI et al. (2015), on the other hand, find high ranges of runoff increase for three time periods and RCP 4.5, the values being much higher in general. On a monthly scale, highest runoff increase can be seen in November, February, and March in both studies and both time periods. It is remarkable that the runoff shows an increasing trend throughout the whole century for RCP 4.5, though the increase is lower until the end of century, thus resulting in a peak of increase in mid-century. Reasons might be a decrease in snow cover area and a lowering of temperature and precipitation increase for the RCP 4.5 scenario and late-century (cf. ALI et al. 2015, p. 14).

Study	Time period	Study area	Findings
Gebre & Ludwig (2014)	2030 & 2070	Indus River outlet	Reference period: 1971-2005; ranges for Nov-Mar: **2030:** + 25% to 75% (RCP 4.5) ; + 75% to 125% (RCP 8.5) **2070:** + 20% to 80% (RCP 4.5) , + 70% to 150% (RCP 8.5) Largest increases predicted for summer (Jun-Aug)
Lutz et al. (2016)	2021 - 2100	UIB	Reference period 1971-2000, ranges for Nov-Mar: **2021 - 2050:** 0 to +25% (RCP 4.5) ; + 40 to 120% (RCP 8.5) **2071 - 2100:** - 20% to + 20% (RCP 4.5) ; + 80% to 140% (RCP 8.5)
Ali et al. (2015)	2006 - 2100	Indus River in the UIB	Reference period: 1976-2005, ranges for Nov-Mar: **2006 - 2035:** + 64% to 218% (RCP 4.5) ; + 73% to 253% (RCP 8.5) **2041 - 2070:** + 85% to 271% (RCP 4.5) ; + 68% to 253% (RCP 8.5) **2071 - 2100:** + 37% to 178% (RCP 4.5) ; + 81% to 576% (RCP 8.5)

Tab. 5: Summary of literature findings on future winter runoff trends for Indus River (own depiction).

In the RCP 8.5 scenario, Indus River runoff shows an increasing trend throughout the whole century, thus not peaking in mid-century. The reasons might be the following: the scenario shows precipitation increase in all seasons except spring, and glacier melt will be accelerated due to major temperature rise. The reduced precipitation in spring is overlayed by the shift of glacier and snow melt from summer to spring, which leads to a major runoff increase in spring as well (cf. LUTZ et al. 2016). ALI et al. (2015) find very high ranges of runoff increase within the cultivation period, again the increase is highest in November, February and March, whereas increase in December and January varies around 75 to 120% throughout the whole century and does not show significant changes between the three time periods. For the results of LUTZ & IMMERZEEL (2016) it is important to note that the range of runoff increase for RCP 8.5 given in Tab. 5 does not differ highly between the two time periods, but the average increase is significantly higher for late-century. For mid-century, only one tributary reaches an increase up to 120%, whilst the others show increases around 40% to 60%. On the other hand, for late-century, the majority of tributaries especially in lower areas shows an increase of 150% from November to January and lower increase in February and March, and tributaries in upper areas show these high values only in March and lower values from 50% to 100% in the other months, so the average increase in late-century is significantly higher (cf. LUTZ & IMMERZEEL 2016).

Since the values differ significantly for the studies and because there are not enough studies used here, mean values for future runoff are not calculated.

3.4 Summary

At this point, future scenarios for mid- and late-century are concluded for mean temperature changes, extreme temperature patterns, Indus River runoff, and the RCP 4.5 and 8.5 scenarios (see Tab. 6). It must be noted that mean temperature increase is given on an annual scale, but runoff increase only for cultivation period. Temperature increase in the winter semester will be a little higher than the annual mean values. Overall, temperature will increase with time and higher emissions, as well as heat waves and hot days, whereas cold or frost days will decrease. Runoff will increase in all scenarios as well, though for late-century and RCP 4.5, a decline of increase will be shown, possibly leading to decreasing runoff compared to the last decades in some winter months. All in all, future winters will be hotter, extreme warm temperatures will occur more often, the number of frost days will decrease, and more water will be available from Indus River.

		RCP 4.5	RCP 8.5
Mid-century	Temp.	+1.1°C (a little higher for Punjab)	+1.5°C (Sindh); +1.6°C (Punjab)
	Runoff	Weak or medium increase (mainly in Nov, Feb, and Mar)	Medium to strong increase (mainly in Nov, Feb, and Mar)
Late-century	Temp.	+2.3°C (Sindh); +2.6°C (Punjab)	+4.5°C (Sindh); +4.8°C (Punjab)
	Runoff	Medium increase (mainly in Nov, Feb, and Mar) or weak decrease	Strong increase
Number of heat waves, (very) hot days will increase, cold / frost days will decrease for both scenarios throughout the whole century			

Tab. 6: Conclusive summary of findings for different scenarios (own depiction).

4 Implications for Wheat Production

After summarizing future climate projections, their implications for wheat production are investigated. To do that, impacts from all three factors are explained individually first because most studies focus on impacts from temperature increase or extreme temperature changes, but do not combine several factors, so this will be done in this study in the end. No study could be found that investigates the impact of runoff changes on wheat production, but some studies make assumptions, and AHMAD et al. (2020) calculated yield change for several crops correlated to water balance, which includes canal deliveries, but also precipitation, evapotranspiration, and groundwater level.

In the past, wheat production has increased in both Sindh and Punjab. Between 2000/01 and 2015/16, the cultivated areas increased by about 10% in both provinces, and yield increased by 21% in Sindh and 14.6% in Punjab. This development is attributed to socio-economic factors like better storing facilities and price policies, but also to "favourable weather conditions"

(KHAN et al. 2021, p. 8000). These findings are supported by AHMED & SCHMITZ (2011), who find an increasing trend in wheat yield as well for 1987 to 2004 (p. 4). This might hint that climate change is already affecting wheat production positively, since runoff and temperature have already increased in the past (see chapters 3.1 and 3.3). However, AHMAD et al. (2014) state that technological improvements play a significant role in yield increase (p. 11).

ABID et al. (2016a) investigate farmers' perceptions of climate risks to crop production in Punjab and find that in average, 42% of all interviewed farmers perceive an uncertainty or decrease of crop production, 27% no change and 15% better production (p. 452). It must be noted, though, that the interviews concentrated on three smaller regions within Punjab, and statements refer to all crops, not only wheat.

It becomes clear that there are many factors that affect wheat production, and precise predictions for future changes can hardly be made. However, implications can be derived from climate change developments and the known optimum temperature values for wheat growth.

4.1 Implications from Mean Temperature Changes

AHMAD et al. (2014) highlight that studies on climate change impact on crop production in Pakistan are contradicting, meaning that some predict adverse and others positive effects, or even no change. Again, these studies use different study areas, variables, and data, which leads to differing results (pp. 2-3). Several studies calculate yield change based on data from the past, though many different factors apart from temperature increase would have to be considered by the models, e.g., fertilization, technological improvement, political situation etc., which most of them do not (cf. ibid., pp. 5-6).

Two main developments can be noticed: in cooler regions like northern Punjab, wheat production is affected positively because optimum temperature is reached more often, and minimum temperature increases as well (cf. ibid., p. 139), whereas in hot regions like Sindh and southern Punjab, temperature increase leads to high maximum temperatures and thus accelerated crop growth so that wheat grains cannot gain proper size and weight (cf. RASUL et al. 2012, p. 93). Conclusively, wheat yield will rather decrease in southern areas (Sindh, southern Punjab), and increase in northern Punjab due to favorable temperatures. Therefore, temperature increase impact on wheat production shows a spatial pattern, though four out of five studies considered here do not make spatial distinctions. Accordingly, the values given in Tab. 7 include climate data for whole Pakistan (AHMED & SCHMITZ 2011; AHMAD et al. 2014) or subregions (MAHMOOD et al. 2021; RASUL et al. 2012), and show a highly differentiated pattern within the study areas. SULTANA et al. (2009) give values for yield change differentiated in arid and semi-arid zones.

Three studies find a wheat yield decrease for increasing temperature (AHMED & SCHMITZ 2011; AHMAD et al. 2014; SULTANA et al. 2009) and one an increase up to a tipping point of 24.3°C (MAHMOOD et al. 2021), whereas AHMAD et al. (2014) find a seasonal pattern with yield increase in January and February and decrease for November and December. SULTANA et al. (2009) find that yield decrease due to temperature increase is higher in semi-arid areas than in arid areas, thus contradicting the assumption above. It must be noticed that their study includes only one meteorological station in the semi-arid zone, but six in the arid zone.

Using the tipping point of 24.3°C from MAHMOOD et al. (2021) and 30°C as a maximum temperature above which wheat growing is affected negatively (see chapter 3), presumptions

Study	Study area	Time period	Findings
Ahmed & Schmitz (2011)	Pakistan	1987 - 2004	Yield change for each 1°C temperature increase: -44kg/ha
Rasul et al. (2012)	Indus Delta (Sindh)	none	Wheat yield decrease because of accelerated crop growth
Sultana et al. (2009)	Sindh & Punjab	none	Reference period: 1971 - 2000 Yield change for each 1°C temperature increase: -2% to -4% (arid zones) -5% (semi-arid zones)
Ahmad et al. (2014)	Pakistan	1981 - 2010	Yield change for each 1°C temperature increase: -7.4% (sowing season, Nov-Dec) +6.4% (vegetative season, Jan-Feb) no significant impact (harvesting season, Mar-Apr)
Mahmood et al. (2021)	Sindh & Punjab	2010 - 2100	Reference period: 1975 - 2005 Yield change for each 1°C temperature increase: +232kg/ha Tipping point: 24.3°C

Tab. 7: Summary of literature findings on impacts of temperature increase on wheat production (own depiction).

can be made from current climate data and future temperature increase. Fig. 4 shows current climate diagrams for Sindh and Punjab, blue lines are added for both key figures as well as the start and end of cultivation period in November and March / April. For Sindh, one can see that mean temperature is already close to the tipping point in November and higher in March and April, and maximum temperature is higher than 30°C in November, and much higher in March and April. This indicates that the temperature in Sindh is already less favorable for wheat production, and a mean temperature increase of 1.1°C (RCP 4.5, mid-century) would deteriorate the situation in sowing and harvesting season, perhaps manifesting by shortened cultivation periods. Indeed, sowing dates in Sindh have been delayed by two to three weeks already in the past years to avoid high temperatures (cf. AHMAD et al. 2014, p. 13). Minimum temperature in

Sindh is above 8°C in all months, but unfavorable temperatures below 5°C might occur, but will probably decrease with temperature rise.

For Punjab, the tipping point is not reached within the cultivation period, and a temperature increase up to 2.6°C (RCP 4.5, late-century) would not be enough for temperature in March and November to reach 24.3°C, but in April. Maximum temperatures are below 30°C in all months

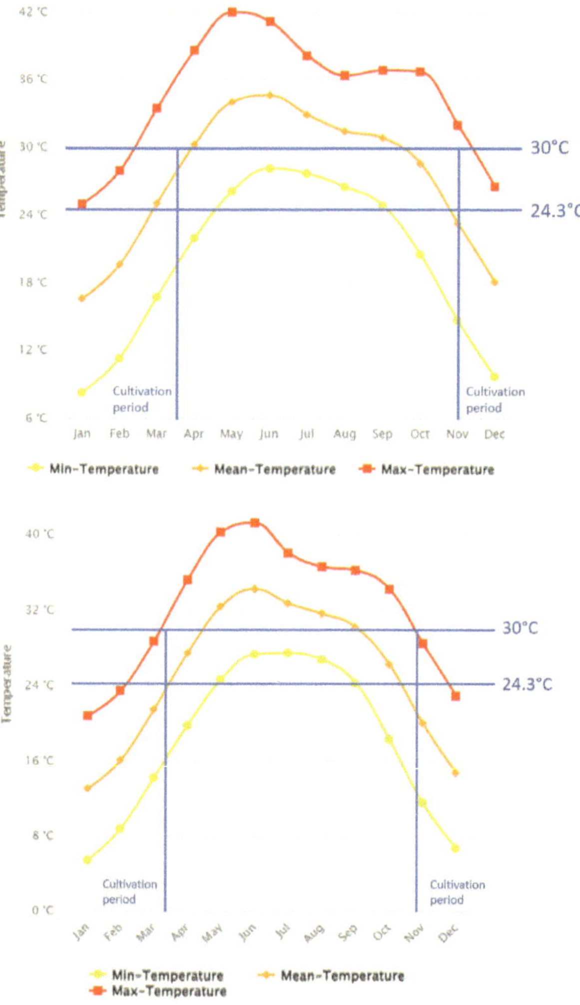

Fig. 4: Monthly min-, mean- and max-temperature in Sindh (top) and Punjab (bottom) from 1991-2020 (World Bank Group 2021, cultivation period, 24.3 °C and 30 °C blue lines added by the author).

as well, meaning that Punjab will show favorable temperature conditions for wheat in the future for medium temperature increase scenarios, because mean temperature will be closer to the optimum and maximum temperatures above 30°C will not be reached too often. On the other hand, minimum temperature is at 6.6°C in December and 5.3°C in January, thus cold temperature has negative impacts on wheat production in Punjab. An increase in minimum temperature would therefore contribute to positive impacts on wheat production.

For both provinces applies that temperature increase in December and January will have positive impacts because mean temperature will be closer to the optimum and minimum temperatures will be higher, but at the same time maximum temperatures are not too high for wheat growth (though temperatures above 30°C will occur more often). For sowing and harvesting season, however, key temperatures are reached faster with temperature increase. This finding corresponds to the seasonal results of AHMAD et al. (2014). For RCP 8.5 scenario and late-century, temperature increase will probably be too high so that wheat production should be affected adversely. In general, wheat production in Punjab will most likely profit from temperature increase even though the increase will be higher than in Sindh, whereas it will probably have negative impacts in Sindh. Here again, though, one must distinguish between arid and semi-arid climate zones within Punjab.

4.2 Implications from Extreme Temperature Patterns

For extreme temperature patterns, no study could be found that calculated yield changes due to increasing numbers of heat waves, hot days or decreasing number of frost days. Several studies indicate that heat waves lead to wheat crop failure (e.g., SAEED et al. 2017; ABBAS et al. 2018), thus, the increase of future winter heat waves will impact wheat production adversely. Since wheat growing stops somewhere above 30°C, the increase of (very) hot days is problematic. Especially Sindh shows a high number of very hot days (T>35°C) and accordingly an even higher number of hot days (T>30°C) during cultivation period, and these numbers will increase in the future mainly in November and March, meaning that the cultivation period will be shortened and wheat growing will stop on several days. In arid and semi-arid areas, yield loss due to shortened crop life cycles is already observed (cf. SULTANA et al. 2009, p. 139).

ABID et al. (2016a) find that most farmers in Punjab are concerned with extreme minimum temperatures in winter because biomass accumulation is reduced (p. 451), and AHMAD et al. (2014) point out that the number of frost days has decreased, but changing temperature patterns occur more uncertainly and frost may occur in late winter, thus having a negative impact on wheat production (p. 13). In the future, the number of frost days will decrease significantly (see chapter 3.2), hence wheat production should be affected positively during cold months. Similar to the seasonal findings in chapter 4.1, the decrease of frost and cold days in December and January will have a positive impact, whereas temperature extreme increase in November, February and March will have negative impacts. The negative effects will be more severe for late-century and RCP 8.5 scenario.

4.3 Implications from Indus River runoff changes

As outlined before, wheat is irrigated with ground water with high share in winter. Snow and glacier melt are still very important as well, though. Fig. 5 shows the share of wheat yield that is attributable to snow and glacier melt. Wheat production in Sindh is more dependent on irrigation from Indus River: snow and glacier melt make an attribution of 50% to 75% in some parts. In Punjab, winter precipitation is higher, and several tributaries are rain-fed (see chapter 3.3), thus, wheat yield is only attributable to snow and glacier melt by 0 to 25% in most areas. This indicates that runoff changes will affect wheat production in Sindh more than in Punjab.

There are no studies that directly correlate runoff changes to wheat production. AHMAD et al. (2020) calculate the effects of different scenarios that also include precipitation and evapotranspiration, but they point out that canal delivery changes have the largest impact on water balance in Sindh (p. 12) and therefore on wheat production. For low flow scenarios, they find that wheat crop area will decrease by 50.000ha to 150.000ha in Sindh, whereas for high flow scenarios, crop area will increase by up to

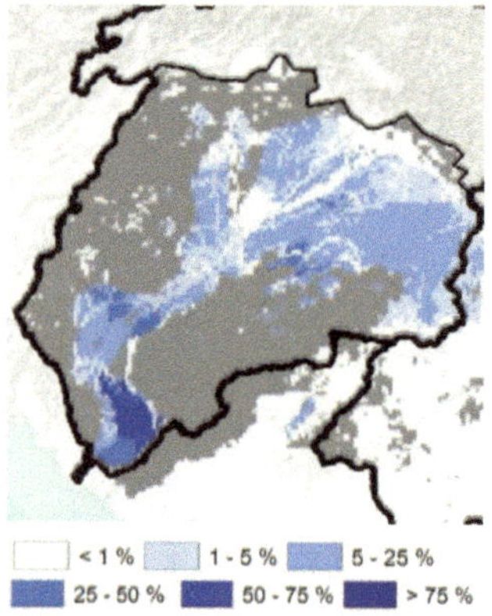

Fig. 5: Wheat yield attributable to snow and glacier melt (BIEMANS et al. 2019).

200.000ha (cf. ibid., p. 34). In this study, a general increase in winter runoff is found (see chapter 3.3), so an increase in wheat crop area according to runoff increase should be observed in the future, and thus perhaps an increase in wheat production. However, AHMAD et al. (2020) concentrate on Sindh only. For northern Punjab, runoff changes might have no significant impact because the restriction to crop area is not as dependent on Indus River as in arid areas, though better water availability tends to affect crop production positively in general. Southern Punjab, being arid and not having water from several tributaries (see Fig. 2, chapter 2), might as well show greater increase in wheat crop area. One must take into consideration that runoff will probably decrease for RCP 4.5 scenario and late-century compared to mid-century, meaning that the increased crop area might decrease again for the end of century. Furthermore, there will be severe water scarcity as soon as there is advanced glacier shrinkage. It must be added that the major runoff increase will happen at the edge of cultivation period in November, February, and March, when there is more water in Indus River anyway. December and January as the months with minimum runoff will show less runoff increase in the future, so it might be necessary to store water from previous months for irrigation in growing phase. Another aspect that must be considered is that with increasing temperature and thus evapotranspiration, the plants' water demand will increase as well.

4.4 Discussion and Assumptions

The effects of temperature increase, extreme temperature patterns and Indus River runoff were assessed individually. Now, general assumptions for future scenarios and all three factors shall be made. Since all changes show a spatial pattern, results are differentiated for Sindh and southern Punjab (arid) and northern Punjab (semi-arid).

Arid areas of Sindh show high mean temperatures and maximum temperatures that are not optimal especially at the beginning and ending of cultivation period. An increase of temperature would shorten cultivation period and sowing dates must be delayed, also wheat growth would be affected negatively because it stops above 30°C. Future extreme temperature events like winter heat waves will increase, meaning additional stress to the crop. Water availability from Indus River will increase, and since this is the major limiting factor to crop area in the arid zone, wheat crop area might be enlarged in the future. These developments combined show that crop yield will perhaps decrease in the future due to temperature increase and heat waves, but the increased crop area might compensate this development, hence the total wheat production might not be affected negatively. However, these are only assumptions because there is no study yet that combines these effects. Since wheat has growing limitations regarding maximum temperature, the high temperature increases for late-century in both RCP scenarios will probably lead to adverse impacts despite major runoff increase, or in addition to the decline in runoff increase, respectively.

Semi-arid areas of Punjab show lower mean temperatures, meaning that the optimum temperature of wheat growth is not reached often, and minimum temperatures below 5°C decelerate wheat growing. On the other hand, maximum temperatures do not pose a problem as they rarely exceed 30°C. Future temperature increase will affect wheat growing positively because mean temperature will be closer to the optimum and minimum temperatures will increase, whereas mean maximum temperatures stay below 30°C in all months except March during cultivation period. Provided that hot days do not occur too often, the overall impact from temperature increase should be positive up to the RCP 4.5 scenario and late-century. For extreme temperature, it is not clear whether the positive impact of the decrease of cold or frost days will exceed the negative impacts from the frost occurrence in late winter. Winter heat waves might have a positive impact in coldest areas within this zone. The runoff increase might have an additional positive effect on wheat production, though it might be minor compared to the effect of temperature increase.

To summarize the findings, general assumptions on wheat production trends can be made as shown in Tab. 8. Positive impacts for semi-arid areas are predicted although the majority of studies predicts rather a decrease of wheat production, but without differentiating their results on a spatial scale. The assumptions made here consider several studies that predict positive impacts for cooler regions. Another restriction is that only areas with irrigation from Indus River can be described, which is only a part of Punjab. Large areas are irrigated with water from tributaries that are more dependent on rain, which are not included in this study.

4.5 Adaptation Strategies

The findings show that adaptation strategies will become more important in the future, especially in Sindh. Adaptation strategies offer the possibility to mitigate temperature increase

	RCP 4.5		RCP 8.5	
	mid-century	late-century	mid-century	late-century
Arid areas	uncertain	negative	negative	negative
Semi-arid areas	positive	positive	positive	negative

Tab. 8: Assumptions for the combined impacts on wheat production from all three factors (own depiction).

impacts on wheat production and to increase irrigation efficiency. Both aspects are briefly examined here.

Temperature increase is accompanied by risks for farmers, but "the degree of their vulnerability depends on their ability to adapt to those risks" (ABID et al. 2016a, p. 454). Early adaptation strategies can make farmers less vulnerable or more profitable than other farmers that adapt lately or do not adapt, respectively (cf. ibid.). ABID et al. (2016b) measure that wheat yield increases significantly with the extent of adaptation strategies and point out that farmers need information about these strategies from the government or non-governmental organizations (p. 263). Strategies might be planting crop varieties, e.g., heat tolerant wheat varieties, replacing vulnerable crop types by others, or changing sowing dates (cf. ABID et al. 2016a, p. 455). However, delaying sowing dates has a rather negative impact on wheat production because the growing season is shortened, which leads to less yield. On the other hand, this measure prevents total crop failure (cf. GORST et al. 2018, p. 24).

Irrigation efficiency is "the fraction of diverted water that reaches the plants and contributes to the crop evapotranspiration" (AHMAD et al. 2020, 21 f.). In Pakistan, this value is quite low, especially in Sindh, because a lot of water is lost through evapotranspiration and seepage in the canals (cf. ibid., p. 22). This implicates that adapting irrigation techniques, perhaps implementing drip irrigation, could increase water availability and thus the crop area. Considering that Pakistan shows a population growth and therefore increasing water demand, and that agriculture is the main water user, water management strategies become important as well. Due to climate change, population growth and urbanization, water demand will increase by around 16% to 25% from 2018 to 2050, hence water management strategies are required to ensure food security (cf. LOHANO & MARI 2020b, p. 404). This means that in the future, wheat producers might compete with other sectors for water from increasing Indus River runoff. Water management strategies for Pakistan have been investigated by many studies (e.g., QURESHI 2011; ULLAH et al. 2020) and contain the increase of water storage capacity or water-use efficiency among others (cf. QURESHI 2011, p. 252).

With respect to these considerations, it is necessary for Pakistan to develop strategies to meet the issues of food security and water scarcity, which will probably be exacerbated by climate change and population growth.

5 Conclusion

This study aimed to derive climate change implications for wheat production in the Lower Indus Basin and has shown that climate change impacts on mean temperature, extreme temperature patterns and Indus River runoff will have different effects on wheat production on a spatial and seasonal scale. For semi-arid areas in northern Punjab, wheat production will probably be affected positively in the future for scenarios up to medium temperature increases, whereas arid areas in southern Punjab and Sindh will more likely show production decrease despite better water availability due to temperature increase. Very high temperatures will be a problem mainly at the beginning and end of cultivation period, which will therefore be shortened especially in arid zones. On the other hand, warmer temperatures in deep winter will affect wheat growing positively especially in cold areas. Winter runoff increases of Indus River will not play a major role in Punjab's wheat production, but in Sindh, resulting in crop area increase. Adaptation strategies are a chance to mitigate negative impacts from climate change. Precipitation changes were not included because their impact on wheat production is non-significant.

There is no study that investigates the impact of temperature increase and runoff trends in combination. Therefore, the aim of this study was to do so by using existing studies on the individual factors, because irrigation plays a major role in the Lower Indus Basin and will be impacted by runoff changes. However, agriculture is a complex system with an interplay of many different factors and stakeholders, so there are many other factors that were not considered such as flooding, pollution, socio-economic or political developments.

In general, the assumptions made in this study base on literature findings that show strongly differing results in some parts. Reasons are not only the different methods used, but also the high number of factors that interrelate with each other and cannot all be considered. Also, predictions for temperature increase vary, but runoff trends vary even more, since they are related to uncertain temperature and precipitation developments. When making statements on effects on wheat production, the studies include all these uncertainties, thus very different results are shown with positive or negative trends, respectively. Additionally, some studies do not make spatial distinctions, although there are different conditions for wheat production due to different climatic zones within the LIB, and climate change manifests differently within the basin. This spatial scale was taken into account in this study.

The findings of this study underline the necessity for agricultural stakeholders in the LIB to adapt to future temperature and runoff changes, if not done already. Since there will be more and more water available from the Indus River, new water storage and distribution systems should be implemented with respect to population growth and growing water demand. Since runoff increase will be lower in wheat growing season (December to February), one should consider storing water in autumn, which shows major runoff increases. Also, it seems to be important for Sindh to test heat tolerant wheat varieties or other methods to mitigate the impact of very hot temperatures. The study has shown that temperature increase and extreme temperature pattern increase will be higher in Punjab, but not having such a negative impact because of a cooler and wetter climate. Since wheat cultivation is concentrated there, wheat production in the LIB should be affected overall positively despite probable decreases in arid areas.

Publication Bibliography

ABBAS, F.; REHMAN, I.; ADREES, M.; IBRAHIM, M.; SALEEM, F.; ALI, S.; RIZWAN, M. & SALIK, M. R. (2018): Prevailing trends of climatic extremes across Indus-Delta of Sindh-Pakistan. In *Theoretical and Applied Climatology* 131, pp. 1101–1117.

ABID, M. (2016): Climate change Impacts and Adaptation in the Agricultural Sector of Pakistan- Socioeconomic and Geographical Dimensions. Hamburg.

ABID, M.; SCHILLING, J.; SCHEFFRAN, J. & ZULFIQAR, F. (2016a): Climate change vulnerability, adaptation and risk perceptions at farm level in Punjab, Pakistan. In *The Science of the total environment* 547, pp. 447–460.

ABID, M.; SCHNEIDER, U. A. & SCHEFFRAN, J. (2016b): Adaptation to climate change and its impacts on food productivity and crop income: Perspectives of farmers in rural Pakistan. In *Journal of Rural Studies* 47 (A), pp. 254–266.

AHMAD, M.; SIFTAIN, H. & IQBAL, M. (2014): Impact of Climate Change On Wheat Productivity In Pakistan: A District Level Analysis. Climate Change Working Paper Series No. 1. Pakistan Institute of Development Economics.

AHMAD, M.; STEWART, J.; PENA ARANCIBIA, J. & KIRBY, M. (2020): Sindh water outlook: Impacts of climate change, dam sedimentation and urban water supply on irrigated agriculture. Technical Report. Sustainable Development Investment Portfolio project. CSIRO.

AHMAD, Z.; HAFEEZ, M. & AHMAD, I. (2012): Hydrology of mountainous areas in the upper Indus Basin, Northern Pakistan with the perspective of climate change. In *Environmental monitoring and assessment* 184, pp. 5255–5274.

AHMED, N. M. & SCHMITZ, M. (2011): Economic assessment of the impact of climate change on the agriculture of Pakistan. In *Business and Economic Horizons* 4 (1), pp. 1–12.

ALI, S.; LI, D.; CONGBIN, F. & KHAN, F. (2015): Twenty first century climatic and hydrological changes over Upper Indus Basin of Himalayan region of Pakistan. In *Environmental Research Letters* 10 (1).

ARCHER, D. R. & FOWLER, H. J. (2004): Spatial and temporal variations in precipation in the Upper Indus Basin, global teleconnections and hydrological implications. In *Hydrology and Earth System Sciences* 8 (1), pp. 47–61.

ARFAN, M.; LUND, J.; HASSAN, D.; SALEEM, M. & AHMAD, A. (2019): Assessment of Spatial and Temporal Flow Variability of the Indus River. In *Resources* 8, article nr. 103, pp. 1–17.

ASGHAR, A.; IQBAL, J.; AMIN, A. & RIBBE, L. (2019): Integrated hydrological modeling for assessment of water demand and supply under socio-economic and IPCC climate change scenarios using WEAP in Central Indus Basin. In *Journal of Water Supply: Research and Technology - AQUA* 68 (2), pp. 136–148.

BIEMANS, H.; SIDERIUS, C.; LUTZ, A. F.; NEPAL, S.; AHMAD, B.; HASSAN, T.; BLOH, W. VON; WIJNGAARD, R. R.; WESTER, P.; SHRESTHA, A. B. & IMMERZEEL, W. W. (2019): Importance of snow and glacier meltwater for agriculture on the Indo-Gangetic Plain. In *Nature Sustainability* 2 (7), pp. 594–601.

GEBRE, S. L. & LUDWIG, F. (2014): Spatial and Temporal Variation of Impacts of Climate Change on the Hydrometeorology of Indus River Basin Using RCPs Scenarios, South East Asia. In *Journal of Earth Science & Climatic Change* 5 (10), pp. 1–7.

GORST, A.; DEHLAVI, A. & GROOM, B. (2018): Crop productivity and adaptation to climate change in Pakistan. In *Environment and Development Economics* 23 (6), pp. 679–701.

HABIB, Z. (2004): Scope for Reallocation of River Waters for Agriculture in the Indus Basin. Paris.

IMRAN, A. & NOUREEN, K. (2021): Wheat Crop Development in Central Punjab (Faisalabad, 2020 – 21). Regional Agrometeorological Centre - Pakistan Meteorological Department. Faisalabad.

IPCC (2019): Climate Change and Land: An IPCC Special Report on climate change, desertification, land degradation, sustainable land management, food security, and greenhouse gas fluxes in terrestrial ecosystems. Edited by P.R Shukla, J. Skea, E. Calvo Buendia, V. Masson-Delmotte, Pörtner, H.-O,., D. C. Roberts et al.

JORGENSEN, D. W.; HARVEY, M. D.; SCHUMM, S. A. & FLAM, L. (1993): Morphology and dynamics of the Indus River: Implications for the Mohen Jo Daro Site. In SHRODER, J. F. (ED.): Himalaya to the sea: Geology, geomorphology and the Quaternary. London and New York: Routledge, pp. 288–326.

KASANG, D. (n. d.): RCP-Szenarien. Hamburger Bildungsserver. Available online at https://bildungsserver.hamburg.de/unsicherheiten-und-szenarien/4105604/rcp-szenarien/, checked on 11/2/2021.

KHAN, F. K. (1991): A Geography of Pakistan: Environment, People and Economy. Oxford.

KHAN, I.; LEI, H.; KHAN, A.; MUHAMMAD, I.; JAVEED, T.; KHAN, A. & HUO, X. (2021): Yield gap analysis of major food crops in Pakistan: prospects for food security. In *Environmental Science and Pollution Research* 28, pp. 7994–8011.

N. KHAN; S. SHAHID; T. B. ISMAIL & X.-J. WANG (2019): Spatial distribution of unidirectional trends in temperature and temperature extremes in Pakistan. In *Theoretical and Applied Climatology* 136, pp. 899–913.

S. KHAN; M. UL-HASAN & M. A. KHAN (2019): Precipitation and Temperature Oscillation and its Effects on the Flow of Indus Water System and Adaptation in the Arid Region, Pakistan (1940-2000). In *International Journal of Economic and Environmental Geology* 10 (2), 79-68.

KHATTAK, M. S.; BABEL, M. S. & SHARIF, M. (2011): Hydro-meteorological trends in the upper Indus River basin in Pakistan. In *Climate Research* 46, pp. 103–119.

LOHANO, H. D. & MARI, F. M. (2020a): Climate Change and Implications for Agriculture Sector in Sindh Province of Pakistan. In *Mehran University Research Journal of Engineering and Technology* 39 (3), pp. 668–677.

LOHANO, H. D. & MARI, F. M. (2020b): Estimating Sectoral Water Demand for Sindh Province of Pakistan. In *Mehran University Research Journal of Engineering and Technology* 39 (2), pp. 398–406.

LUTZ, A. F. & IMMERZEEL, W. (2016): Reference climate dataset for the Indus, Ganges, and Brahmaputra River Basin. Working Paper. Himalayan Adaptation, Water and Resilience.

LUTZ, A. F.; IMMERZEEL, W. W.; KRAAIJENBRINK, P. D. A.; SHRESTHA, A. B. & BIERKENS, M. F. P. (2016): Climate Change Impacts on the Upper Indus Hydrology: Sources, Shifts and Extremes. In *PloS one* 11 (11).

MAHMOOD, F.; KHOKHAR, M. F. & MAHMOOD, Z. (2021): Investigating the tipping point of crop productivity induced by changing climatic variables. In *Environmental Science and Pollution Research* 28, pp. 2923–2933.

MAHMOOD, H. Z. (2009): Resource Distribution and Productivity Analysis within Pakistan's Agriculture: A Case Study. Berlin.

NEPAL, S. & SHRESTHA, A. B. (2015): Impact of climate change on the hydrological regime of the Indus, Ganges and Brahmaputra river basins: a review of the literature. In *International Journal of Water Resources Development* 31 (2), pp. 201–218.

POMEE, M. S. & HERTIG, E. (2021): Precipitation projections over the Indus River Basin of Pakistan for the 21st century using a statistical downscaling framework. In *International Journal of Climatology*, pp. 1–26.

PORTER, J. R. & GAWITH, M. (1999): Temperatures and the growth and development of wheat: a review. In *European Journal of Agronomy* 10 (1), pp. 23–36.

QURESHI, A. S. (2011): Water Management in the Indus Basin in Pakistan: Challenges and Opportunities. In *Mountain Research and Development* 31 (3), pp. 252–260.

RAJBHANDARI, R.; SHRESTHA, A. B.; KULKARNI, A.; PATWARDHAN, S. K. & BAJRACHARYA, S. R. (2015): Projected changes in climate over the Indus river basin using a high resolution regional climate model (PRECIS). In *Climate Dynamics* 44, pp. 339–357.

RASUL, G.; MAHMOOD, A.; SADIQ, A. & KHAN, S. (2012): Vulnerability of the Indus Delta to Climate Change in Pakistan. In *Pakistan Journal of Meteorology* 8 (16), pp. 89–107.

SAEED, F.; ALMAZROUI, M.; ISLAM, N. & KHAN, M. S. (2017): Intensification of future heat waves in Pakistan: a study using CORDEX regional climate models ensemble. In *Natural Hazards* 87, pp. 1635–1647.

SAJJAD, H. & GHAFFAR, A. (2019): Observed, simulated and projected extreme climate indices over Pakistan in changing climate. In *Theoretical and Applied Climatology* 137, pp. 255–281.

SHAFQAT, M. N.; Maqbool; A.S.; Eqani, S. A. M. A. S.; AHMED, R. & AHMED, H. (2016): Trends of Climate Change in the Lower Indus Basin Region of Pakistan: Future Implications for Agriculture. In *International Journal of Climate Change Strategies and Management*.

SULTANA, H.; ALI, N.; IQBAL, M. M. & KHAN, A. M. (2009): Vulnerability and adaptability of wheat production in different climatic zones of Pakistan under climate change scenarios. In *Climatic Change* 94, pp. 123–142.

ULLAH, H.; AKBAR, M. & KHAN, F. (2020): Construction of homogeneous climatic regions by combining cluster analysis and L-moment approach on the basis of Reconnaissance Drought Index for Pakistan. In *International Journal of Climatology* 40 (1), pp. 324–341.

World Bank Group (2021): Climate Change Knowledge Portal. Country Pakistan. Available online at https://climateknowledgeportal.worldbank.org/country/pakistan, checked on 10/28/2021.

YU, W.; YANG, Y.-C.; SAVITSKY, A.; ALFORD, D.; BROWN, C.; WESCOAT, J.; DEBOWICZ, D. & ROBINSON, S. (2013): The Indus Basin of Pakistan. The Impacts of Climate Risks on Water and Agriculture. Washington D.C.

YOUR KNOWLEDGE HAS VALUE

- We will publish your bachelor's and master's thesis, essays and papers

- Your own eBook and book - sold worldwide in all relevant shops

- Earn money with each sale

Upload your text at www.GRIN.com
and publish for free